Nature Numbers

# HOW SLOW IS A SLOTH?

## Measure the Rain Forest

Jill Esbaum

Children's Press®
An imprint of Scholastic Inc.

Library of Congress Cataloging-in-Publication Data
Names: Esbaum, Jill, author.
Title: How slow is a sloth? : measure the rainforest / Jill Esbaum.
Description: New York : Children's Press, and imprint of Scholastic Inc., [2022]. | Series: Nature numbers | Includes index. | Audience: Ages 5-7. | Audience: Grades K-1. | Summary: "Nonfiction, full-color photos of animals and nature introduce basic math concepts and encourage kids to see a world of numbers all around them"– Provided by publisher.
Identifiers: LCCN 2021031693 (print) | LCCN 2021031694 (ebook) | ISBN 9781338765212 (library binding) | ISBN 9781338765229 (paperback) | ISBN 9781338765236 (ebk)
Subjects: LCSH: Measurement–Juvenile literature. | Rain forests–Miscellanea–Juvenile literature. | Animals–Miscellanea–Juvenile literature. | BISAC: JUVENILE NONFICTION / General | JUVENILE NONFICTION / Mathematics / General
Classification: LCC QA465 .E83 2022 (print) | LCC QA465 (ebook) | DDC 530.8–dc23
LC record available at https://lccn.loc.gov/2021031693
LC ebook record available at https://lccn.loc.gov/2021031694



10 9 8 7 6 5 4 3 2 1 22 23 24 25 26

Printed in the U.S.A. 113
First edition, 2022

Series produced by WonderLab Group, LLC
Book design by Moduza Design
Photo editing by Annette Kiesow
Educational consulting by Leigh Hamilton
Copyediting by Jane Sunderland
Proofreading by Molly Reid
Indexing by Connie Binder

Photos ©: back cover frog: Isselee/Dreamstime; cover: Michael S. Nolan/Science Source; 1 frog: Isselee/Dreamstime; 3: Edurivero/Dreamstime; 4-5: KenCanning/Getty Images; 6-7: Dirk Ercken/Alamy Images; 7 inset: Brandon Alms/Dreamstime; 8 inset center: Brandon Alms/Dreamstime; 10-11: Milosk50/Dreamstime; 11 inset: Brandon Alms/Dreamstime; 14-15: Feathercollector/Dreamstime; 16: Hotshotsworldwide/Dreamstime; 18-19: Wim van den Heever/Nature Picture Library; 21: Martin Mcmillan/Dreamstime; 22-23: Andamanse/Dreamstime; 24-25: Pete Oxford/Minden Pictures; 26-27: Thierry Montford/age fotostock; 31 frog: Isselee/Dreamstime; 32: Brandon Alms/Dreamstime.

All other photos © Shutterstock.

For Bria, Lawson, and Leo
–JE

Try It!

Look for answers to all the "Try It!" panels on page 31.

# Hello, Sloth.

Hope you don't mind
if we watch you climb.
Step one: Plant those
feet and reach up, up, up.
Step two: Pull in your
**s-l-o-w m-o-t-i-o-n** way.
Repeat.

Take your time, Sloth.
We’ll check back later.

poison dart frog

## Shiny skin, sticky toes.

While leaping from leaf to leaf, a poison dart frog often stops to look around.

The bright skin of these tiny frogs warns other animals, "Don't eat me, I'm poisonous!"

### Try It!

This frog is 1 inch (2.5 cm) long. If 14 poison dart frogs of the same size sat in a row, how long would that row be?

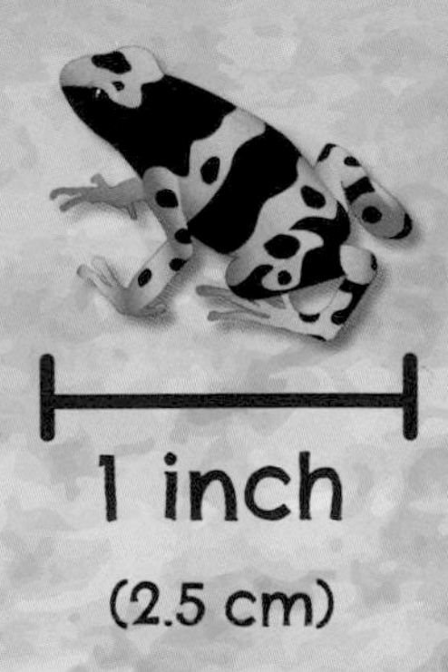

Tree boas slide silently from tree to tree, searching for food. This snake is just a baby. But already it can curl its tail around branches and hold tight.

## Try It!

These animals are different sizes.

baby tree boa = 12 inches (30 cm)

frog = 1 inch (2.5 cm)

sloth = 20 inches (50 cm)

Put these animals in order from biggest to smallest. Then try smallest to biggest.

baby tree boa

jaguar

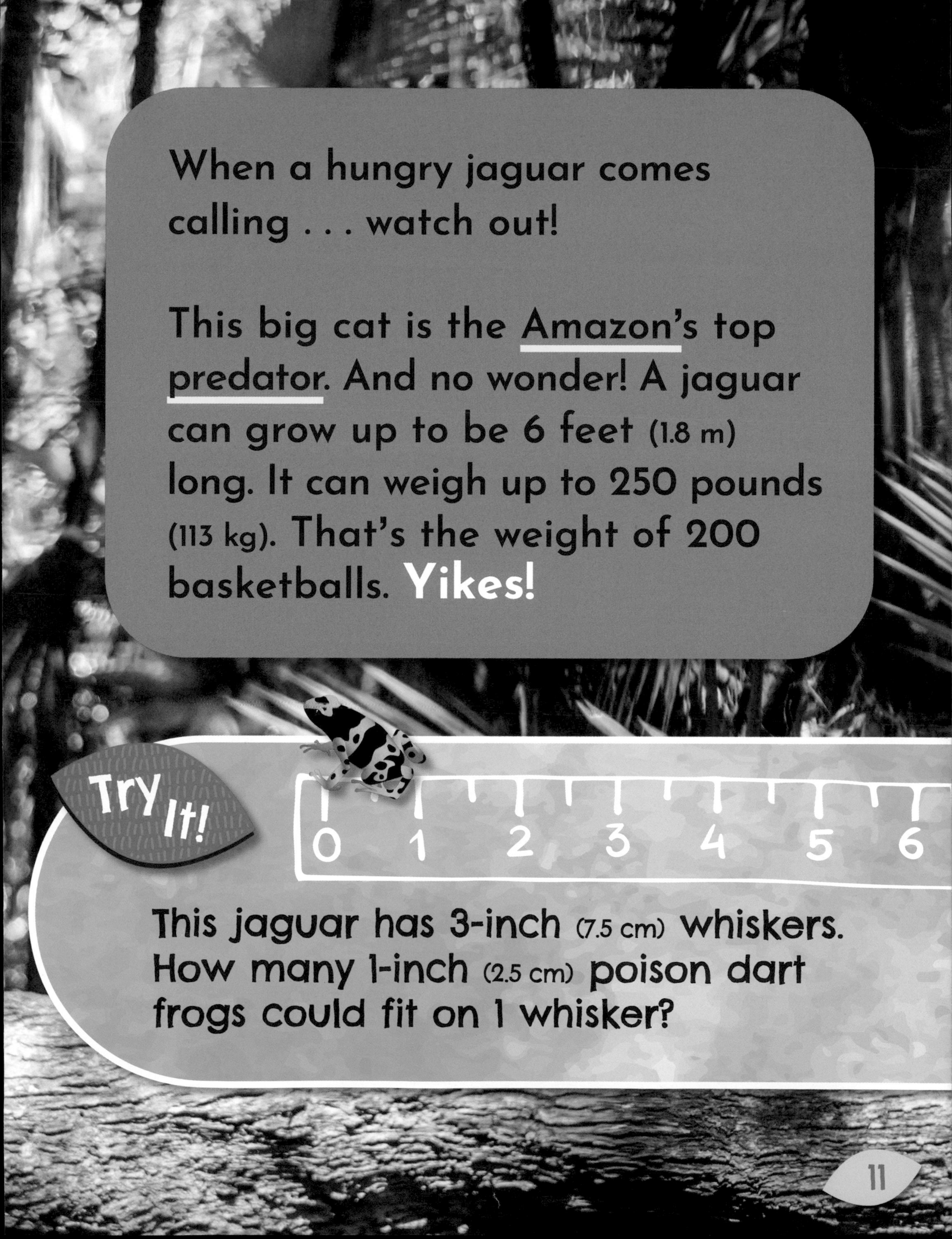

When a hungry jaguar comes calling . . . watch out!

This big cat is the Amazon's top predator. And no wonder! A jaguar can grow up to be 6 feet (1.8 m) long. It can weigh up to 250 pounds (113 kg). That's the weight of 200 basketballs. **Yikes!**

**Try It!**



This jaguar has 3-inch (7.5 cm) whiskers. How many 1-inch (2.5 cm) poison dart frogs could fit on 1 whisker?

Hoatzins (wah-ZEENS) have spiky hairdos. Their feathers are sunset colors, and their faces are blue. They group together on low branches.

Predators give them plenty of space. Why? Their nickname holds a clue: stinkbird. Quick, hold your nose!

A bird sat on a branch at 8:00 a.m. Four more birds arrived 30 minutes later. What time was it then?

hoatzin

**Each morning,**
this beetle crawls from its leafy hideaway to seek a fruit snack.

The Hercules beetle, the strongest and largest in the world, has a body that grows to 3 inches (7.5 cm) long. But check out its awesome horns. Those add another 4 inches (10 cm)!

## Try It!

**Which is longer, the beetle's body or its horns?**

body length = 3 inches (7.5 cm)

horn length = 4 inches (10 cm)

Hercules beetle

# Hello again, Sloth.

Taking a break?
No wonder. Climbing
is hard work.

At least those 3-inch-long (7.5 cm) claws will make it easy to hang from a high branch.

Go. Do your thing, Sloth . . .
Slow and steady.

## Try It!

The sloth started climbing at 9:00 a.m. Now it is 3 hours later. What time is it?

Capybaras (ka-pub-BAR-uhs) are the largest rodents in the world. They spend much of their time in streams, grazing and eating. Capybaras will bark to warn others if they see a jaguar . . . **or a sneaky caiman.**

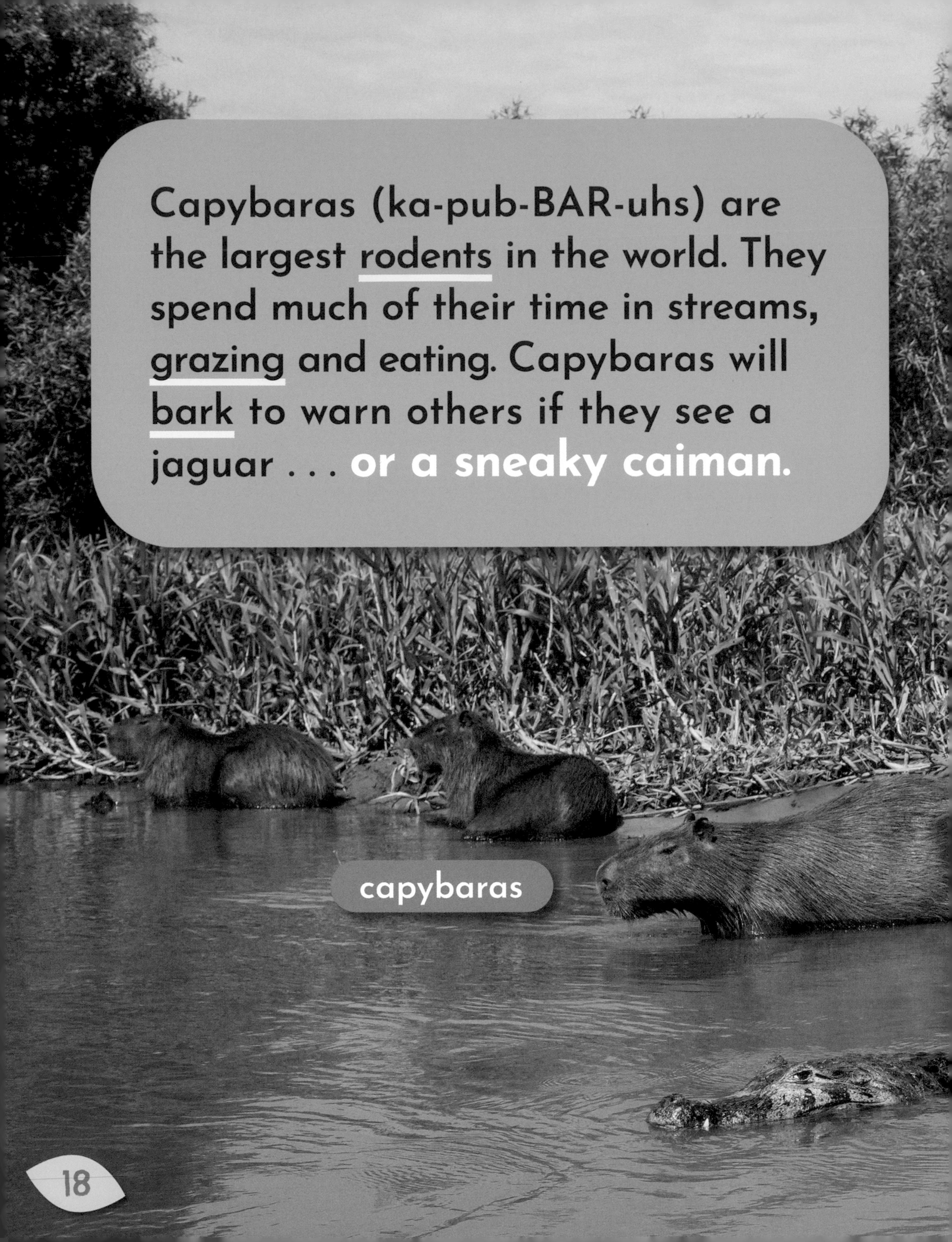

capybaras

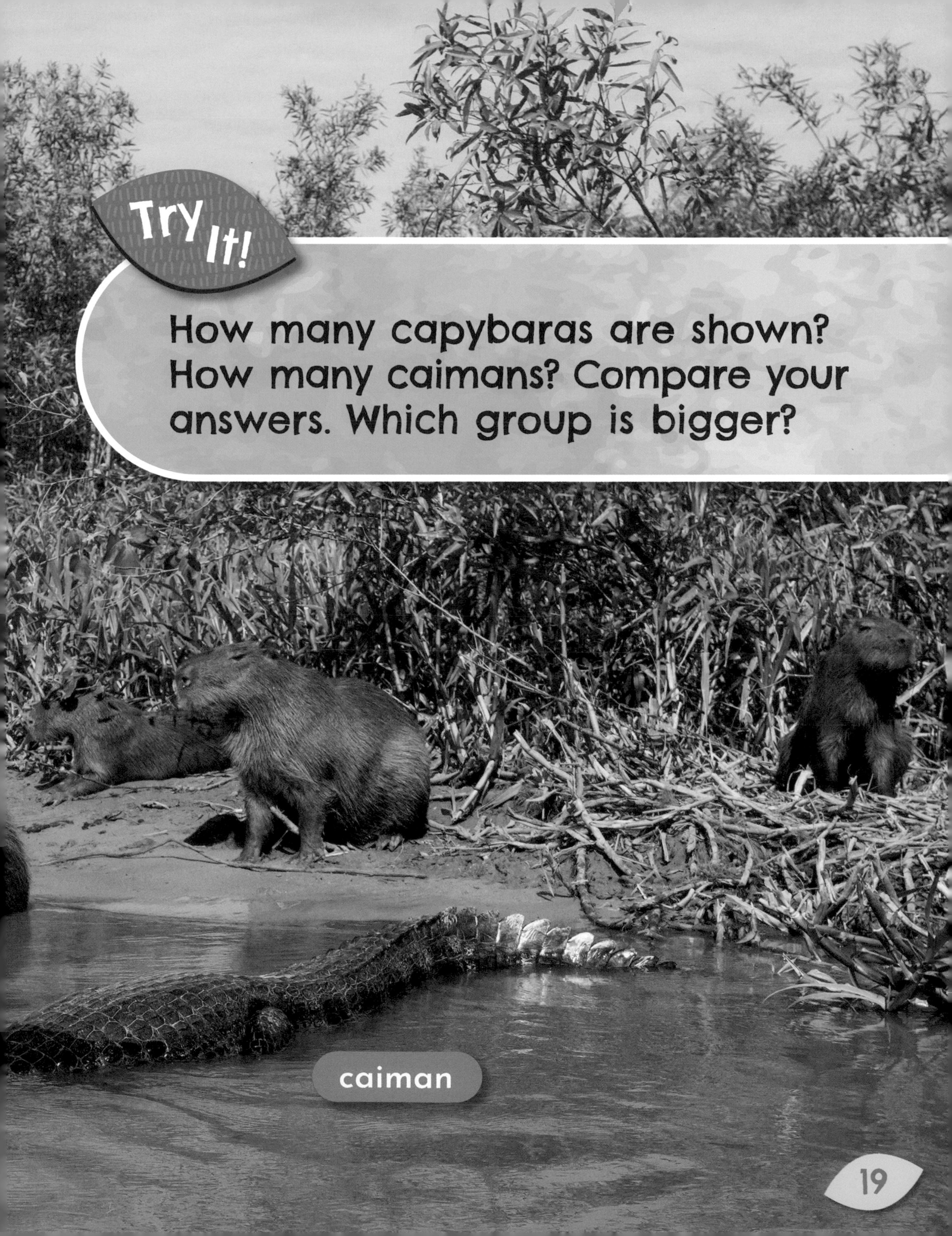


## Try It!

How many capybaras are shown? How many caimans? Compare your answers. Which group is bigger?

A tamarin has a small, easy-to-read face. A grumpy stare tells us we are too close. **Oops.**

Tamarins swing from branches and vines. They often stop for a bite of ripe fruit . . . or an unlucky lizard. Their long fingers grab tasty bugs from trees. **Yum!**

**Tamarins are small monkeys.**

tamarin = 8 inches (20 cm)

1 foot = 12 inches (30 cm)

**Is this monkey more or less than 1 foot (30 cm) tall?**

tamarin

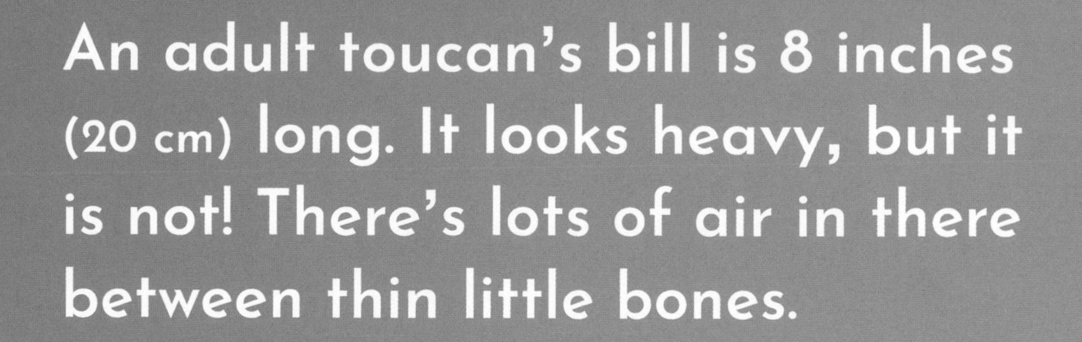

An adult toucan's bill is 8 inches (20 cm) long. It looks heavy, but it is not! There's lots of air in there between thin little bones.

Toucans use their bills like tools to reach a berry up high or to peel an orange.

## Try It!

A toucan chick's bill is much shorter than an adult's bill.

chick's bill = 2 inches (5 cm)

adult's bill = 8 inches (20 cm)

How many inches (centimeters) does the chick's bill have left to grow?

Turtles paddle the muddy rivers and quiet ponds of the Amazon rain forest. They eat water plants, fallen fruit, and seeds.

When feeling chilly, they warm up in the tropical sunshine.

Each of these turtles is about 1 foot (30 cm) wide.

5 turtles = 5 feet wide (152 cm)

If 2 turtles swam away, what would the last 3 turtles measure, all together?

Arrau turtles

# You made it, Sloth!

## Good climbing.

So . . . how slow is a sloth, the slowest mammal on Earth? That depends. On the ground, a sloth moves about 10 feet (3 m) per minute. If you raced a sloth, you could walk very slowly and still win!

But a sloth ***can*** move faster if it feels afraid, up to 15 feet (4.5 m) per minute.

# Bye for now, Sloth.

Nice hanging out with you awhile!

# Activities

In this book, readers practiced measuring Amazon animals, adding, subtracting, and comparing. Here are measuring activities kids can do to practice their skills.

Read these activities to the kids and help them take the fun beyond the pages of this book.

## SLOW AS A SLOTH (pages 4–5, 16–17, 26–27)

When a sloth is on the ground, it pulls itself along by reaching forward with its front feet, digging in its claws, then dragging itself ahead. Try crawling like a sloth. Have somebody time you to see how far you can move in one minute.

## LITTLE LEAPERS (pages 6–7)

Poison dart frogs leap from leaf to leaf looking for food. Measure how far you can leap from a taped line on a floor or sidewalk. Try it two ways: standing still and running up to the line before jumping.

## JAGUAR JUMP (pages 10–11)

A jaguar has such powerful back legs it can leap 10 feet (3 m) straight up into a tree. Have a friend or family member measure your vertical jump (straight up from standing still). Then you measure your friend's jump. Who jumped higher?

## FOUND POUNDS (pages 12–13)

Stinkbirds weigh about 2 pounds (1 kg). If you have a scale, weigh some items from around your house. What weighs more than 2 pounds? What weighs less?

## AS LONG AS A CAPYBARA (pages 18–19)

A capybara is about 4 feet (1.2 m) long. This length can be measured in many different ways using different items. Gather bunches of like items from around your home—things like crayons, paper clips, spoons, or even pillows. Mark a 4-foot-long (1.2 m) space on the floor with tape, then lay your items into that space end-to-end. How many of each item does it take to add up to 4 feet (1.2 m)?

## ONE, TWO, AS LONG AS A SHOE (pages 20–21)

Tamarin monkeys are only about as long as your shoe. What else can you find in your house that is about as long as one of your shoes? What is as long as two of your shoes?

## COUNTING CATCHES (pages 22–23)

Toucans often toss fruit to one another using their bills. Place two chairs facing each other a few feet apart. You sit in one and have a friend or family member sit in the other. Play catch with a small ball using only one hand. Count as you catch. How many catches can you make before missing? If it is too easy, move the chairs apart a little at a time.

## TURTLE TRAIL (pages 24–25)

When baby Arrau turtles hatch from their shells, they are about 2 inches (5 cm) long. Measure your arm from shoulder to fingertip. How many Arrau hatchlings could line up in that space?

# Glossary

**Amazon** (AM-uh-zon) The world's largest tropical rain forest, located in South America.

**bark** (bahrk) A sudden harsh sound in the throat.

**claw** (klaw) A hard, sharp nail on the foot of an animal or a bird.

**feather** (feTH-ur) One of the light, soft parts that cover a bird's body.

**grazing** (GRAY-zing) Feeding on grasses.

**grumpy** (GRUHM-pee) Easily irritated; grouchy.

**poisonous** (POI-zuh-nuhs) Having a poison that can harm or kill.

**predator** (PRED-uh-tur) An animal that lives by hunting other animals for food.

**rodent** (ROH-duhnt) A mammal with large, sharp front teeth that are constantly growing and used for gnawing things.

**spiky** (SPY-kee) Something formed into sharp points.

**steady** (STED-ee) Uniform and continuous.

**tropical** (TRAH-pi-kuhl) Of or having to do with the hot, rainy area of the tropics.

## Answers

**PAGE 7** 14 inches (35 cm)

**PAGE 8** 1. sloth, baby tree boa, frog 2. frog, baby tree boa, sloth

**PAGE 11** 3 frogs

**PAGE 12**
8:30 a.m.

**PAGE 15** the horns

**PAGE 17**
12:00 p.m. or noon

**PAGE 19**
6 capybaras, 1 caiman; the group of capybaras is bigger.

**PAGE 20** less

**PAGE 23** 6 inches (15 cm)

**PAGE 25** 3 feet (91 cm)

# Index

Page numbers in **bold** indicate illustrations.



# ABOUT THE AUTHOR

Jill Esbaum lives on an Iowa farm and is the author of nearly fifty books for kids. She writes picture books, including *Where'd My Jo Go?*, *We Love Babies!*, *How to Grow a Dinosaur*, and *If a T. Rex Crashes Your Birthday Party*; the Thunder and Cluck series of graphic early readers; and nonfiction books about nature, history, and famous people.